Global Evolution of Dual-Use Biotechnology

A Report of the Project on Technology Futures and Global Power, Wealth, and Conflict

Project Director
Anne G.K. Solomon

Author
Gerald L. Epstein

April 2005

About CSIS

The Center for Strategic and International Studies (CSIS) is a nonprofit, bipartisan public policy organization established in 1962 to provide strategic insights and practical policy solutions to decisionmakers concerned with global security. Over the years, it has grown to be one of the largest organizations of its kind, with a staff of some 200 employees, including more than 120 analysts working to address the changing dynamics of international security across the globe.

CSIS is organized around three broad program areas, which together enable it to offer truly integrated insights and solutions to the challenges of global security. First, CSIS addresses the new drivers of global security, with programs on the international financial and economic system, foreign assistance, energy security, technology, biotechnology, demographic change, the HIV/AIDS pandemic, and governance. Second, CSIS also possesses one of America's most comprehensive programs on U.S. and international security, proposing reforms to U.S. defense organization, policy, force structure, and its industrial and technology base and offering solutions to the challenges of proliferation, transnational terrorism, homeland security, and post-conflict reconstruction. Third, CSIS is the only institution of its kind with resident experts on all the world's major populated geographic regions.

CSIS was founded four decades ago by David M. Abshire and Admiral Arleigh Burke. Former U.S. senator Sam Nunn became chairman of the CSIS Board of Trustees in 1999, and since April 2000, John J. Hamre has led CSIS as president and chief executive officer.

Headquartered in downtown Washington, D.C., CSIS is a private, tax-exempt, 501(c) 3 institution. CSIS does not take specific policy positions; accordingly, all views expressed herein should be understood to be solely those of the author(s).

Library of Congress Cataloging-in-Publication Data
Epstein, Gerald Lewis.
  Global evolution of dual-use biotechnology : a report of the project on technology futures and global power, wealth, and conflict / Gerald L. Epstein.
      p. ; cm.
  ISBN 0-89206-459-5 (alk. paper)
  1. Biotechnology—Social aspects. 2. Biotechnology—Government policy.
  [DNLM: 1. Biotechnology—trends—Congresses. 2. Security Measures—Congresses. TP 248.23 E64g 2005] I. Title.

  TP248.23.E64 2005
  303.48'3—dc22

                                        2004031112

The CSIS Press
Center for Strategic and International Studies
1800 K Street, N.W., Washington, D.C. 20006
Tel: (202) 887-0200
Fax: (202) 775-3199
E-mail: books@csis.org
Web: http://www.csis.org/

Contents

Preface

Throughout 2004, at the request of the National Intelligence Council (NIC),[1] the Center for Strategic and International Studies (CSIS) brought together leaders from advanced technology firms, venture capital enterprises, research universities, and government to consider the geopolitical, economic, and social implications of technological advance out to the year 2020. This publication is one of a series that reflect their deliberations.[2]

The purpose of the exercise was to provide the NIC with ideas and insights relevant to NIC 2020, a project designed to provide U.S. policymakers with a view of how world developments could evolve and to identify opportunities and potentially negative developments that might warrant policy action. The challenge for CSIS was to produce for the NIC analyses that would be useful over time in the face of an explosion of scientific and technological knowledge and with great uncertainties in their paths of growth and patterns of interaction with economic, social, and other forces.

CSIS work included a series of workshops, conferences, and commissioned papers that considered frontier research and innovation and related policy challenges: the global ubiquity of powerful, dual-use technologies; the increasing diversity of players in the global advanced technology enterprise that operate beyond the influence of governments or international governing bodies; and the policy dilemma of balancing security-driven imperatives to curtail the availability of dual-use assets with the need of advanced technology enterprise for openness.

Numerous individuals provided advice and assistance throughout the project. Adrianne George, CSIS program coordinator, provided extensive intellectual and administrative assistance. We owe special debt of gratitude to Dr. Charles A. Sanders, CSIS trustee and project chair, and to Frank C, Carlucci, honorary cochair. CSIS would also like to thank Ambassador Robert L. Hutchings, former chairman of the National Intelligence Council, for his encouragement and support.

Anne G.K. Solomon
Project Director

---

[1] The NIC is a center of mid-term and long-term strategic thinking within the U.S. government, reporting to the director of central intelligence and providing the president and senior policymakers with analyses of foreign policy issues.

[2] Sandra Braman, *Information Technology, National Identity, and Social Cohesion* (CSIS, 2005); Anthony J. Cavalieri, *Biotechnology and Agriculture in 2020* (CSIS, 2005); Gerald L. Epstein, *Global Evolution of Dual-Use Biotechnology* (CSIS, 2005); Julie E. Fischer, *Dual-Use Technologies: Inexorable Progress, Inseparable Peril* (CSIS, 2005); David Nagel, *Wireless Sensor Systems and Networks: Technologies, Applications, Implications, and Impacts* (CSIS, 2005); and Anne G.K. Solomon, ed., *Technology Futures and Global Power, Wealth, and Conflict* (CSIS, 2005).

Global Evolution of Dual-Use Biotechnology

Gerald L. Epstein
Senior Fellow
Center for Strategic and International Studies

This report draws on presentations and discussions at a workshop convened by the Center for Strategic and International Studies on March 18, 2004, as part of the "Technology Futures and Global Power, Wealth, and Conflict" project supported by the National Intelligence Council.

Executive Summary

On March 18, 2004, the Center for Strategic and International Studies held a workshop to explore the global evolution of bioscience and biotechnology and to examine the security implications of their inherently dual-use natures. Presentations and discussion among the roughly 40 participants, from the industrial, academic, analytic, and policymaking communities, addressed some of the field's key characteristics and their implications. The main conclusions include:

Biotechnology offers a set of powerful tools whose capabilities are rapidly expanding. The range of techniques, processes, and applications encompassed by the term "biotechnology" is very broad. By any measure, however, the field is growing rapidly, both in extent and in capability. Scientific publications reflect our exponentially expanding ability to understand and manipulate the processes of life. Genetic information from an increasing number of species is being deciphered and stored, and we are improving our ability to analyze and understand it. And this scientific information is increasingly being put to practical application, whether to inform the development of medical therapeutics or therapies, to tailor organisms to have specific desirable properties or functions, or to take advantage of or emulate specific biological processes to produce materials or equipment.

As one example, "directed-evolution" techniques can very rapidly produce molecules or organisms with desired properties, even if the origins of those properties are not understood. Biotechnology-based diagnostic tools might be able to covertly reveal unprecedented amounts of information about people and where they have been. And the genetic engineering approaches that allow the creation of "designer organisms" might someday be applied to human beings—a possibility that must be considered, whether or not western societies would choose to pursue it.

Biotechnology is globally ubiquitous. Biological science and biomedicine have long been internationally collaborative endeavors—a characteristic that has persisted over time. Advanced capabilities in bioscience and biotechnology can today be found all over the world. Dissemination of the technology is enhanced by the relatively low cost of the necessary materials, equipment, and facilities. Suppliers of equipment, materials, and technical services are found all over the world, meaning that activities that once would have required an extensive and highly trained in-house staff can now be performed with a smaller, less-skilled staff using commercially available equipment and supplies, or can be contracted out entirely.

Competitive market pressures induce firms and research institutions to seek technically capable partners in developing countries, where they can be found at much lower cost and in looser regulatory environments. Indeed, the relatively low cost and the worldwide availability of talent have encouraged a number of developing-world governments to invest in biotechnology as a promising area in which to compete. Consequently, biotechnology is likely to be a leveling factor across the world, rather than one in which the developed states enhance their economic advantage. For all these reasons, the United States and other developed nations no longer lead in all areas of bioscience and biotechnology—if they ever did—and where their leads persist, they will diminish over time.

Biotechnology is inherently dual-use and poses serious security concerns. The characteristics that make biotechnology such a powerful source of medical, scientific, industrial, and other legitimate applications offer equivalent power to those who deliberately seek to inflict harm—a classic example of a "dual-use" technology that can be used for beneficial or for hostile purposes. For example, the ability to produce organisms or materials with specific properties does not distinguish between legitimate and malicious applications. Indeed, causing a disease is easier than curing it. Therefore, the development and geographic spread of bioscience and biotechnology pose serious security concerns, even as they offer tremendous benefits.

These security concerns grow as the technology proliferates in several senses of that word, referring to expansion not only of the knowledge base and capability provided by the technology, but also of the number of people with the ability to exploit that knowledge base and the number of regions in the world where that expertise can be found. Taken together, these trends point toward a qualitatively new form of security threat—one that, in its limiting case, could provide single individuals with the ability to develop and deploy biological weapons that have serious regional, national, or even global consequences. In such a nightmare scenario, the capability to commit mass destruction could end up in the hands of people whose motives might not even be comprehensible, let alone predictable or deterrable.

Malicious applications will be difficult to detect or control. Biotechnology's pervasive dual-use nature and global dissemination means that malicious activities will take place amidst a much larger set of fully legitimate yet technically quite-similar activities, and hence will be hard to detect. They will also be hard to control, since they typically do not require distinctive or expensive facilities, expertise, or equipment. Difficulties with imposing controls also stem from the widespread dissemination of the technology to date; the diversity of suppliers of equipment, reagents, materials, and support services that would have to be part of any control scheme; the expanding base of legitimate applications and desire not to interfere with them; and the ability of malefactors to work around procedural barriers that may be imposed to block potentially dangerous activities.

The rapid growth of scientific and technological capability also presents the prospect of technological surprise. In legitimate applications, surprises are called innovations, and they can lead to the creation of highly beneficial products and services. In hostile applications, however, technological surprises can negate preventive or protective measures and can therefore constitute serious threats.

Possible control mechanisms might involve regulating materials, equipment, information, or personnel. However, given the proliferation of the technology and expertise and the continued lowering of technical barriers, few such measures seem likely to succeed against determined attempts to subvert, negate, or evade them. Moreover, such measures would necessarily also affect legitimate activities—activities that will not only improve health, welfare, and standards of living, but that must also be relied on to defend against the malicious use of the same technology. Therefore, there are limits to the extent to which any control measures can be imposed.

Can defenses outpace offenses? What will be the outcome of a competition between the beneficial and hostile applications of bioscience and biotechnology, or between the ability of malefactors to develop advanced new weapons and that of defenders to counter them? The attacker appears to have an easier job than the defender for at least three reasons. Two apply to any dual-use technology: first, breaking something is usually far easier and far cheaper than fixing it; and second, the attacker can select a specific time, place, and method of attack, while defenders must prepare for any eventuality. Third, and more specific to biotechnology, is the fact that defenders must prepare medical countermeasures within an extensive regulatory regime designed to ensure safety and efficacy, whereas attackers are not concerned about legal compliance and may not care about their own welfare. This asymmetry offers attackers a flexibility and agility that will be difficult for defenders to match.

Ultimately, however, we have no way of knowing the extent to which malicious applications of bioscience and biotechnology will actually be pursued by individuals, groups, or nations. Theoretical threats do not necessarily translate into real ones. The probability that an individual, a terrorist group, or a nation will mount a biological attack depends on the nature of the technologies that are available both to it and to its intended victims, but that probability likely depends even more strongly on motivations and intentions that are unrelated to any particular technological choice. Moreover, the fact that offense may be easier than defense might be mitigated by the fact that defenders have far greater resources at their disposal.

Introduction

On March 18, 2004, the Center for Strategic and International Studies convened a workshop on the global evolution of biotechnology and the security implications of its inherently dual-use nature. The workshop was part of a major new CSIS project, "Technology Futures and Global Power, Wealth, and Conflict," which explored the economic, social, and security implications of technological frontiers. The National Intelligence Council supported this project as part of its NIC 2020 Project.

The workshop featured presentations on advances in bioscience and biotechnology; the diffusion, decentralization, and "commodification" of biotechnology; governance approaches to biotechnology; and implications for biodefense. Although a prime motivation was to explore potential security threats posed by emerging bioscience and biotechnology, the workshop was not limited solely to that topic.

Workshop participants were invited from a range of disciplines. All quotations in this report not referenced to other publications are from workshop participants.

Biotechnology: Some Emerging and Future Prospects

Although no attempt was made to be comprehensive, a number of current and prospective developments in biotechnology were described during workshop sessions:

- Genetically engineered humans,
- High-throughput cellular and molecular evolution,
- Single-molecule detection and "genetic snooping,"
- Insertion, manipulation, and expression of foreign genes.

Genetically Engineered Humans

According to Charles Cantor, chief scientific officer at Sequenom, "during the next 20 years we will see genetically engineered humans begin to populate this planet." Representing what he described as the next stage of human evolution, such a development will be the result of the maturation of the technical tools with which to perform such manipulation; the desire to identify genetic origins and explore genetic therapies for a number of complex diseases such as diabetes and cancer; and the existence of regions of the world that would tolerate or encourage human genetic modification. "Such actions are viewed in Western and Oriental cultures very differently," he added.

We already have a wealth of experience engineering simpler organisms, Cantor explained. For example, we can do "anything we want" with a mouse, having engineered "tens, maybe hundreds of thousands" of novel mouse species. It has not happened yet in humans because the technology has not yet been perfected, and the cost of failure along the way is not seen as acceptable: "If I engineer mice and one survives and 10 die, I'm perfectly happy." However, Cantor does not believe that the application of these techniques to humans is very far off. Although this was not a consensus view—another participant thought that the prospect of genetically engineered humans in this way was neither as likely nor as imminent as Cantor implied—developments in genomics, along with the quest to develop therapies for complex diseases that have so far resisted treatment, will create pressure to explore human genetic engineering.

Cantor pointed out that diseases such as various forms of cancer and schizophrenia are not amenable to conventional drugs, nor do they appear to be amenable to treatment with more-complex antibody therapies. Research to date has identified genetic origins to many such diseases, and the therapies to address these diseases will involve correcting or interfering with the function of the responsible genes. Although the type of gene therapy that is likely to be seen in the near term would not make changes in human reproductive cells and would therefore not be passed to subsequent generations, it nevertheless "pushes the envelope towards human evolution" in which germ-line, or reproductive, cells would be modified. "Germ line is the stuff that is really worrisome," according to Cantor, "and it's permanent."

A diversity of attitudes regarding stem-cell research—research on cells derived from human embryos that may have the potential to develop therapies for a wide range of diseases—may be mirrored with respect to germ-line human engineering. Cantor recounted having heard a public speech in which former Chinese president Jiang Zemin asserted that China would genetically engineer its population in the hope of containing medical costs. Furthermore, the desire to develop humans that can regenerate limbs or organs, or that have enhanced sensory capabilities, may prove irresistible. Should such practices become adopted in certain places or by certain individuals, whether openly or covertly, society will have to deal with the inevitable inequities of living in a world where some have been engineered and some have not.

Finally, Cantor reminded the workshop that engineered human evolution need not be slow. Dr. Göran Sandberg, a distinguished Swedish plant genetic engineer, has developed poplar trees that flower in 16 weeks rather than 16 years, and similar approaches might be applicable to human development.

High-Throughput Molecular/Cellular Evolution and Screening

Current biotechnological techniques that have the potential to rapidly create a wide range of new microorganisms and molecules will prove to be very influential. Rather than individually testing compounds or organisms for desired properties, these techniques work by generating a huge and diverse set of candidates and screening them all simultaneously. The key to this approach is developing a mechanism to select the organism or molecule with the desired properties out of a much larger population of those that do not. Compared to the traditional method of individually testing candidates, this massively parallel approach features:

- Great improvements in productivity,

- Vast increases in ability to create desired properties in molecules and organisms,

- The ability to produce new molecules or cells with the desired properties in a matter of days.

"DNA Snooping"

Evolving analytical techniques will be able to detect and identify trace amounts of DNA that people inevitably leave behind wherever they have been. These techniques are extensions of existing polymerase chain-reaction (PCR) approaches, which are highly specific methods of detecting and reproducing segments of DNA that contain specified genetic sequences. If the DNA collected in a sample and put through such a test contains the target genetic sequences, called primers, those sequences are duplicated; if it does not contain the primers, nothing happens. Each time this procedure is repeated, the amount of DNA matching the primers will double, and eventually it will accumulate to the point that it becomes detectable. If primer sequences are selected that appear in the DNA being searched for, but nowhere else, this approach constitutes a very sensitive and selective test for the target DNA.

This kind of test can be susceptible to "false positives" if the primer sequences that had been thought to uniquely identify the DNA of interest actually occur in other types of DNA. In such cases, developing an approach that requires a third or a fourth primer to be present at the same time can reduce the number of false positives essentially to zero.

Similarly, techniques that could sequence very small amounts of DNA (i.e., determine the entire set of their genetic codes) could also be used to provide highly specific, essentially error-free confirmation of whether DNA with a specified set of codes was present. Such technology could be used initially in medical diagnostics to detect DNA known to be associated with pathogens. However, it could also be used for monitoring personnel. "I'm willing to bet that 10 years from now, when you go to the airport to get on your plane," said Cantor, "a genetic test will be done to confirm your identity. That's pretty hard to fake."

Insertion, Manipulation, and Controlled Expression of Foreign Genes

These techniques, which are already standard biotechnological practices, have important implications for the future. Robert Erwin, chairman of the board of Large Scale Biology Corporation, described work being done in two small businesses, one of which—Large Scale Biology in Vacaville, California—inserts genes for proteins of interest (such as those for a particular vaccine) into plant viruses that have been modified to extend the range of host species that they can infect. According to Erwin, this firm has been able to engineer viruses that deliver biologically active peptides and proteins and has scaled the process up to produce several tons of virus or viral products.

The second company—Icon Genetics in Germany—takes genes for products of interest from nonviral organisms and splices them into viruses that will integrate themselves directly into the genome of the organisms they infect. In their new location, these viruses remain dormant until activated by specific chemical or biological triggers. The net result, according to Erwin, is "to facilitate control of the viruses for purposes of gene expression in ways that do not occur in nature and in fact never have occurred in nature."

The Evolution of Biotechnology: Changes over Time

Bioscience and biotechnology are developing at a phenomenal pace as more people and firms work in the field; more papers are published; and more products, processes, and techniques are developed. According to Robert Carlson, research scientist in the Department of Electrical Engineering at University of Washington, "what you can do in the lab is changing with extraordinary speed."

Carlson presented several quantitative measures that indicate the rate of progress in this field.[1]

1. The productivity with which DNA can be sequenced (i.e., the order determined of the individual elements that comprise the genetic code) or synthesized (i.e., assembled so that its genetic code has a predetermined sequence) is increasing with time at a rate that is considerably faster than the famous Moore's Law, which describes how rapidly the ability to manufacture microelectronic circuits has improved over time.[2] That is, the

---

[1] This material is elaborated in Robert Carlson, "The Pace and Proliferation of Biological Technologies," *Biosecurity and Bioterrorism*, vol. 1, no. 3 (2003): 203–214.

[2] Moore's Law states that the number of transistors or circuit components on a microchip doubles about every 18 months to two years, a rate that has persisted over several decades and across several major shifts in microelectronics-production technology.

amount of DNA that one individual can sequence or synthesize per day, or per year, is increasing dramatically with time.

2. The cost per base pair for sequencing or synthesizing DNA has dropped accordingly.

3. The time required to determine the structure of a protein molecule—an experimentally complex and computationally intensive task—has similarly dropped, and the productivity with which researchers can do that work (structures obtained per researcher per year) has increased. Whereas the first protein structures took years to work out, they can now be determined in days or even hours, and the time required continues to decrease.

This rate of progress—and the inevitable surprises and breakthroughs—makes prediction particularly difficult. "Why would we think we would be really good at predicting biology between now and 2020?" asked one participant. "In fact, when we look at our track record with respect to it, we've been horrible. The people who made major discoveries in biology in substantial measure didn't even know what they had, frequently, for several years after they discovered it."

At the same time, some workshop participants believed that the rapid rate of progress in biological science and technology did not necessarily correspond to equivalently rapid changes in society or the economy. Although science and technology have indeed advanced rapidly, the translation of these new technologies into new products in the marketplace, especially in the case of approved drugs and therapeutics, has never kept up. For the last 15 years, according to Kris Venkat, chairman of Morphochem, Inc., "the pharmaceutical industry has tried very hard to adapt every imaginable new technology to speed up and find new drugs, but by and large we have failed." Similarly, another participant commented that getting a product through research and exploratory development is very different than getting it into the marketplace, given the steep regulatory hurdles involved.

And despite the rate of progress in bioscience and biotechnology, we do not yet have the ability to do engineering design in these disciplines in the same sense that that term is used in other disciplines. With the notable exception of software engineering, modern engineering is based on the principle of "predictive design": modeling techniques, analytic procedures, testing equipment and methodologies, and an extensive base of empirical experience that permit the performance of engineered systems to be reliably predicted before they are built. In such an environment, engineers can be held accountable for establishing and meeting professional standards.

As Robert Carlson pointed out, we do not yet have quantitative and predictive design tools for engineering organisms or systems of molecules, and we have only a handful of "composable parts" (i.e., standardized components that can be selected and assembled in various combinations to achieve useful results). As a consequence, we are poorly equipped to predict, or handle, surprises.

The Evolution of Biotechnology: Dissemination across Countries and Markets

Affordability and Geographic Spread

Biotechnical expertise, facilities, and supporting infrastructures such as supplier firms have become thoroughly globalized, extending to the developing world as well as throughout the industrialized nations. This worldwide dissemination has been driven by a number of factors. Historically, people around the world have had a common interest in health issues. In addition to any altruistic motivations they may have had, scientists and public-health officials have been motivated to work together because diseases spread; outbreaks in one part of the world can disseminate as fast and as far as people can travel. As a result, extensive and high-quality international collaborations in biomedical and biological science go back for decades, and they have spurred the development of relevant expertise around the world. Robert Erwin, who has built a number of research collaborations involving scientists from countries such as Russia and Ukraine, stated that "we found very smart people from all of those countries, and in hiring we tried to get the best people we could." Collaborations were also driven by economic considerations, he said, since "I knew I could hire people or contract out work [in these countries] at a much lower cost than I could if I did it in England or France or the United States."

Even in the developed world, equipment and materials costs in many areas of biotechnology can be quite low, particularly when compared to those in other scientific and technological disciplines. Expensive, highly specialized, and capital-intensive facilities are not typically required. "It costs a few hundred dollars to hack together a GMO [genetically modified organism] at this point," according to Robert Carlson, and key tools are cheap and readily available. "$50,000 and a trip to e-Bay and the hardware store will buy you any lab you like for manipulating biological things."

The relatively low cost and the worldwide availability of talent have contributed to biotechnology's rapid spread around the world, a spread that is being actively encouraged by a number of developing-world governments, which see it as a promising area in which to compete. "Far from causing greater inequities, inequalities, I think we'll see [biotechnology] produce wealth and produce changes…relatively evenly throughout the world," said Carlson. The United States and other developed nations do not lead in all areas today, and in those areas where they are leading, that lead will erode over time.

Worldwide Growth of Suppliers

Associated with the worldwide spread of biotechnical expertise and research activity is a corresponding dissemination of supporting infrastructure. "There are little companies in virtually every part of the world now that supply various aspects of biotechnology reagents and supplies," explained Erwin.

Market Penetration

As various applications of the technology have manifested themselves, biotechnology-based goods have assumed increasing economic importance. Carlson cited figures showing biotechnology to be responsible today for some one-third to one-half of 1 percent of the U.S. gross domestic product, with a significant number of firms (close to one-fifth of those included in a recent U.S. government survey) engaged in industrial or agricultural, rather than medical, applications.[3] In a similar vein, a McKinsey survey shows that biotechnology has played some role in the production of 5 percent of the world's chemical sales, with the potential to grow to 20 percent by 2010.[4] Market penetration continues overseas as well; Kris Venkat relayed that more than half the cotton acreage planted in countries such as India, China, and Brazil is genetically modified.

Adding current and future medical applications back in, and recognizing that the field is currently dominated by R&D investment as opposed to production, there is great potential for expansion. After all, a large fraction of current biotech firms "are 5 to 15 years away from even a fantasy of a hope of making a profit," as one of the workshop participants put it, and it is their future potential, rather than current earnings, that keeps their investors interested.[5]

National Regulatory Issues

Biotechnology's spread into commercial markets around the world is heavily influenced by national legal and regulatory systems. In discussions of the rate at which new biotechnology applications were being discovered in the lab, several workshop participants said that any such development in the health-care or agricultural arenas would have to surmount major, and time-consuming, regulatory hurdles before attaining commercialization and significant market penetration. Accordingly, differences between national regulatory systems can influence those countries' relative rates of progress in these areas. On the other hand, industrial applications of biotechnology face no such regulatory hurdles, and for that reason it may be important to disaggregate this type of application from the more heavily regulated ones in future analyses.

National differences with respect to gene therapy, and its ultimate application in the production of genetically engineered humans, have been mentioned above, and it is interesting to note that the world's first licensed gene therapy is in China, not the United States or Europe.[6] Attitudes regarding research on stem cells

---

[3] Data from a U.S. Commerce Department survey, as described in Alexandra Goho, "U.S. Government Previews Biotech Survey," *Nature Biotechnology*, vol. 21, no. 8 (August 2003): 837–838.

[4] "Field of Dreams," *The Economist*, April 10, 2004.

[5] At the same time, many biotech firms are actually making money. Forty percent of the firms included in a recent U.S. Commerce Department survey of U.S. biotechnology activities were breaking even or making a profit. Goho, "U.S. Government Previews Biotech Survey": 837–838.

[6] In October 2003, China licensed a form of gene therapy for regular clinical use in the treatment of head and neck cancers. David Cyranoski, "From SARS to the Stars," *Nature*, December 18/25, 2003.

obtained from unwanted human embryos also reveal marked national differences, with governments such as Singapore eager to attract research that the United States at present does not permit to be done with government funds and may deem to be illegal in any form. Similarly, Robert Erwin said that delays in winning approval to conduct clinical trials of new pharmaceuticals (including new biotechnology products) in the United States are driving costs up to the point that these trials are being outsourced internationally.

However, the sense that regulatory differences would persist over time, with the least-restrictive national regulatory system leading to long-term competitive advantage, was not universally held. At least one participant suggested that in the long run, the regulatory regimes around the world would converge.

International Regulatory Issues

One treaty that could affect the deployment of biotechnology is the Biodiversity Convention[7] and in particular its Biosafety Protocol. According to one participant, this treaty, which among other things governs the shipment of living genetically modified organisms across national borders, has not as yet affected the development of biotechnology in the laboratory, but it could constrain the worldwide spread and adoption in the marketplace of certain biotechnology products. If not implemented rationally, it could provide a significant impediment and disincentive to future R&D efforts as well.

Security Threats Posed by Advances in Biotechnology

Much of the workshop addressed biotechnology's dual-use potential: the technology has great power to benefit mankind, but the very same expertise and procedures can be used to inflict great harm. In fact, as Charles Cantor noted and others agreed, "causing a disease is probably easier than preventing or curing disease."

In light of biotechnology's dual-use nature, the National Academy of Sciences' National Research Council convened a Committee on Research Standards and Practices to Prevent the Destructive Application of Biotechnology in 2002. Charged to explore measures to impede biotechnology's malicious applications without interfering with its legitimate ones, this committee began a 2003 report by acknowledging "the capacity for advanced biological research activities to cause disruption or harm, potentially on a catastrophic scale."[8]

---

[7] Committee on Biological Diversity, concluded at Rio de Janeiro, Brazil, on June 5, 1992.

[8] Committee on Research Standards and Practices to Prevent the Destructive Application of Biotechnology, National Research Council, *Biotechnology Research in an Age of Terrorism* (Washington, D.C.: National Academies Press, 2003). The committee that produced this report is also called the Fink Committee after its chair, Gerald Fink, from MIT's Whitehead Institute for Biomedical Research. The quotation referred to appears on page 1.

The characteristics that were described in the previous section—worldwide diffusion; modest capital requirements; and the profusion of supplies and service firms—make biotechnology accessible to those who may wish to do harm. As Robert Erwin explained,

> You don't have to be a state-sponsored program [to manipulate viruses maliciously]. You don't even have to be a large corporate program. You can literally be a very small group of people with a very modest amount of resources, and you can buy all of this stuff in catalogs and on the Internet and through the word-of-mouth network that anybody trained in these fields has.

Multidimensional Proliferation

The potential for biotechnology to be used for harm is proliferating in all the meanings of that term: the underlying knowledge base is expanding, and with it the possibility of technical surprise; the number of people with the technical ability to exploit that knowledge is increasing, due to the lowering of costs and other barriers; and the number of geographic regions of the world where people can be found with the requisite technical ability is also growing. As Erwin stated, "everything will be cheaper, faster, and in the hands of a vastly larger number of people who are competent to use it." In such an environment, it is difficult to predict either the nature of the threat that these technologies might make possible, or where that threat might be coming from. "There's lots of stuff out there coming down the road that I can't even guess at," said Robert Carlson, "and that's what concerns me. We don't have the capability to really recognize it until something significant has happened, nor do we have the capability to respond to it."

Some Specific Concerns

Given these characteristics, biotechnology could manifest or exacerbate a number of security concerns that include:

- *Genetic-activation technology.* As described in a previous section, technologies are in use today that allow foreign genes to be engineered into viruses, integrated into the DNA of animals or people infected by those viruses, and then activated by a subsequent chemical, biological, or environmental trigger.

- *Ease of production scale-up.* Modern production facilities are capable of producing mass quantities of biotechnology-derived substances or organisms. One commercial firm using viruses to perform genetic engineering on plants had, at one point, produced enough genetically modified tobacco mosaic virus to inoculate the entire southern tobacco belt in the United States.

- *"Reload."* Richard Danzig, Sam Nunn Prize Fellow at CSIS and former secretary of the navy, stated that anyone who would be able to produce biological agents of sufficient quality and quantity to conduct a major attack would likely be able to produce much more than would be needed for a single attack. Therefore, authorities would have to expect that any successful bioterrorist might quickly "reload" and attack again.[9] This inability to predict the ultimate scale of a bioterrorist attack would greatly complicate the ability to allocate resources to manage such an incident and would substantially exacerbate public concern.

- *Accessibility of the technology.* Charles Cantor pointed out that state-of-the-art techniques are not needed to wreak devastating harm and that a modest level of capability could suffice: "There are some really scary scenarios that you can write down. There are some really powerful tools…[that] allow you to do things in very short periods of time, and they can be done in a garage. They're not at all sophisticated. They're sophisticated intellectually, but they're not sophisticated technologically."

- *Novel modes of attack.* Of course, it is not just the availability of these technologies that gives rise to concern, but also the nature of the harm that might be done with them. Without wanting to be too specific in responding to a question about how bad a bioattack could be, Cantor warned of some particularly ominous possibilities: "I can imagine scenarios where people with malicious intent could construct just absolutely terrifying things. Things, if you want to imagine this, far worse than just killing everybody. So we have to be very careful about this."

- *Individual empowerment.* A qualitatively new category of security concern manifests itself in the limit where a single person can develop and deploy weapons that can have extremely serious consequences. Workshop participants did not attempt to estimate how much damage a single person could do, either now or in the future, but they were cognizant of trends in technology that are putting more and more power in the hands of smaller and smaller groups. People acting on their own are not held back by the need to find others who share their views, nor are they vulnerable to exposure by collaborators who leak their plans. Perhaps most ominous is the possibility that a lone individual could be so psychopathic or irrational that he or she might seek to commit mass destruction for reasons that could not be anticipated or explained, much less deterred. As Danzig put it, "Optimism with respect to deterrence or control here is, I think, difficult to sustain when you start talking about individuals…because individuals are ultimately more numerous than any other groups, and because they're more erratic—and therefore ultimately the probability that one does something really crazy is quite discomforting."

---

[9] This idea is elaborated in Richard Danzig, "Catastrophic Bioterrorism—What is to be Done?" (Washington, D.C.: Center for Technology and National Security Policy, National Defense University, August 2003).

Difficulty in Detecting and Controlling Malicious Applications of Biotechnology

Biotechnology's pervasive dual-use nature does not just mean that the capability for malicious use is widely disseminated. It also means that those malicious uses will take place amidst a much larger set of fully legitimate yet technically quite-similar activities and hence will be very hard to detect. Malicious activities will also be very hard to control, since neither they nor their legitimate counterparts typically require distinctive, specialized, or expensive facilities, materials, expertise, or equipment.

Difficulty in Predicting or Detecting Malicious Activity

The fact that otherwise naturally occurring disease agents can be used for intentional attacks complicates the ability to detect and identify malicious acts. One participant stated that if an outbreak of foot and mouth disease were to occur in the United States, it would be very hard to determine whether it was of natural or terrorist origin. In such a case, "the government is not going to look very credible."

The small "footprint" of malicious biotechnical activity makes it particularly hard to identify. Genetic engineering can be performed in ways that do not leave any distinctive "signature" in the organisms that have been modified; similarly, the type of molecular- and cellular-evolution work described above to generate substances or organisms with particular properties would also be very difficult to detect or to trace. The fundamental problem, as expressed by David Franz, chief biological scientist at the Midwest Research Institute, is that "intent is the key, and we don't have an intent meter we can put on people."

Predicting where the next threat may emerge is a similar challenge. The possibility of technical surprise—the discovery of a new technique or principle that could be used for harm, but that had not or could not have been predicted—is always present. In addition, one participant noted that deliberate misuse of biology comprises only a portion of the future biological threat. According to this participant, the National Academy study cited above (the Fink Committee report) was probably more concerned about the inadvertent contribution that legitimate scientific work might make to weapons activities than it was about the deliberate misuse of science, since those directly responsible for developing biological weapons are not likely to be much influenced by a National Academy recommendation. Furthermore, nature alone—or more accurately, the interaction between nature and human activity—is fully capable of unleashing an outbreak of an emerging infectious disease, such as SARS, or a reemerging one, such as the pandemic flu that swept the planet early in the last century.

Difficulty in Controlling Malicious Activity

With a range of possible adverse outcomes having been laid before them, workshop participants devoted considerable attention to the prospects for controlling the malicious use of biology. Unfortunately, there are a number of reasons that make that a challenge:

- *It's too late.* The relevant technologies and expertise have spread to the point where it would be impossible to limit further work to a restricted environment. According to Kris Venkat, "I don't think you can do it. It is bound to fail." Techniques that had some value in stemming nuclear proliferation, including tight controls over key materials and classification not only of bomb-design details but also some of the fundamental underlying science, will not work in biology. According to Robert Erwin, "it's too cheap, it's too fast, there are too many people who know too many things, and it's too late to stop it."

- *Pathogens and information are readily available.* Despite the attention being given to controlling pathogens considered to be particularly useful as weapons—so-called select agents—Robert Erwin pointed out that there are lots of readily available viruses that are not select agents but that could be converted to human pathogens by virologists who know how to isolate and manipulate them. Moreover, the genetic-sequence information that would be needed to do this kind of engineering is publicly available—and with advances in sequencing technology, it wouldn't even have to be. Erwin explained that a well-funded terrorist operation could generate its own genomic data.

- *Supporting resources are widely available.* One consequence of the worldwide availability of suppliers for various key materials, reagents, and services is that those pursuing malicious applications need not do everything on their own. In principle, the use of external suppliers might provide an audit trail that investigators could use to learn more about illicit programs. However, according to Robert Erwin, "there are so many different suppliers [for certain types of reagent] that it would be virtually impossible to get a list of all of the customers."

- *There are too many ways around procedural barriers.* Procedural barriers to impede weapons-development activities do not appear promising. As desirable as putting "speed bumps" in the way of potential terrorist applications might seem, said Erwin, in practice they would be "completely irrelevant. It's too easy to engineer around them." He added that in an area where he had personal expertise, a "small team of virologists in a brainstorming exercise could easily engineer around all of the current surveillance safeguards and sort-of bureaucratic approaches to preventing the engineering of very effective pathogens." Ultimately, "not only are there multiple ways to get to a given biological end point, but there are a lot of different people that supply the various materials to get there." Robert Carlson concluded that "the only people speed bumps would slow down is us…. We're not in control of the direction of research at this point around the world."

Mitigating Factors

Will the Real Threat Be as Bad as the Imagined One?

Although much of the workshop discussion painted a fairly dire picture of the future bioweapons threat, some participants offered a more optimistic view. One cautioned against overreacting to hypothesized worst cases—a sentiment echoed by another who observed that, at various times in the past, then-emerging technologies such as nuclear weapons were projected to lead to similarly dark outcomes. Those worst-case scenarios never materialized, and we should not assume that today's "chicken little" equivalents will either. A third participant said, "I think one has to really believe in the basic morality of humankind."

One mitigating factor would be the extent to which terrorists really sought to wreak the greatest possible destruction. Arguing that they would not, one participant stated that what terrorists usually seek is confidence that their attacks will succeed. They "tend to use conservative technology that they know will work" and that will be effective in inspiring fear, rather than aspiring to globally catastrophic consequences.

A second mitigating factor would be the extent to which even an engineered pathogen could put huge numbers of people at risk. One participant said that human diversity means that human immune responses to a disease would be widely variable and that therefore "there will be people immune to any given disease." If true, such a circumstance would relegate to science fiction the concern that a novel biological agent would sweep the planet, killing all in its path. However, another participant countered that although humans are genetically diverse in some ways, they are not so far apart in most—and assuming that diversity will serve to protect in this manner would be a dangerous foundation on which to base public policy.

Moreover, it remains hard to dismiss worst-case scenarios once it is recognized that bioterror threats may not come just from groups seeking to use terror to attain some political or religious objective, but that in the long run, psychopathic individuals may pose significant threats as well. The problem would not be an individual who is so disturbed that he or she would be incapable of planning and executing an attack, but rather one who is highly capable but had a different set of motivations and self-constraints—a "biological Unabomber," as Richard Danzig put it. According to Robert Erwin,

> The problem is permanent. There will be bio hackers. There will be little disenfranchised groups that have the knowledge and the education and the skill to buy these modules [i.e., the functional expertise, equipment, and/or materials needed to engineer pathogens] and put them together. And, unfortunately, out of the billions of people on the planet, there will be a few who have the perverse motives who would like to see that kind of an outcome.

Since there is little prospect of being able to deny terrorists or sociopaths access to technologies, materials, resources, and information sufficient to produce devastating consequences, the case for optimism essentially rests on one of three possibilities:

- that the threat will not, in fact, materialize;

- that social responses will be able to mitigate offensive applications of the technology; or

- that defensive applications of bioscience and biotechnology can be developed and deployed sufficiently rapidly to counter the threat.

While the first possibility may turn out to be the case, it would seem difficult as a matter of policy to have great confidence in that prospect, especially considering the potential consequences of being wrong. The second possibility was not discussed much at the workshop—but if individuals or groups are committed to do harm and have access to the means to do so, social responses face quite a challenge. In the long run, therefore, society's hopes appear to rest on the third possibility, which was the subject of considerable discussion.

Can Defenses Run Faster than Offenses?

Unfortunately, workshop participants were fairly pessimistic about a defensive solution. "Why do you think the defense can outrun the offense?" Elisa Harris, senior research scholar at the University of Maryland, asked after a participant had stressed the importance of defensive efforts. "That's certainly not the experience in history with any other technology."

Indeed, there are a number of asymmetries that favor the offense in such a competition. As Robert Erwin explained, offense is easier. "It's so much easier to destroy things than it is to fix them." Moreover, attackers have a tremendous range of options at their disposal. They are able not only to choose the agent, target, dissemination means, and time and place of attack, but they are able to do so with a fair degree of knowledge about the extent and nature of our open society's preparations and responses. Defenses, on the other hand, have great difficulty in covering the range of possibilities. Although medical countermeasures such as therapeutics and vaccines can and are being developed, it is very difficult to develop "broad-spectrum" defenses that can counter an equivalently wide range of attack modes.

If therapies or antidotes could be developed and deployed in "near-real time," they would not have to anticipate every possible attack, but could rather respond to the specific biological agents that had actually been used. However, such a system would not only face enormous technical hurdles in developing and producing countermeasures, but would also have to overcome nontechnical and noncommercial difficulties such as rapidly recognizing the attack, deciding how to respond, and disseminating the appropriate countermeasures on a mass scale. One participant was pessimistic about overcoming such nontechnical issues, even if the very difficult technical ones could be solved.

Medical countermeasures, unlike offenses, are developed within an extensive regulatory environment to ensure they are safe, effective, and cost efficient. According to Robert Erwin, "the defense is embroiled in a very lethargic, large system of bureaucracy, whereas the offense is extremely mobile, extremely flexible, and unconstrained by any of the very important concerns about safety, risk management, public health, and all of that." Offenses also find value in testing, stated Richard Danzig, but testing "can be effected simply by the device of attacking and attacking again and again and seeing what happens."

One advantage that defensive activity has over offensive activity is that of scale. The legitimate scientific enterprise is vastly larger than the organizations that may be applying these technologies for harm. Therefore, some of the asymmetries that favor the offense might be ameliorated by sheer quantity. Moreover, legitimate work—unlike weapons development—need not keep all evidence of its existence a secret.

One complicating factor, however, is the possibility that defensive efforts can stimulate the very offensive actions they are trying to mitigate—a problem that gets worse as the magnitude of the defensive effort increases. Programs to develop defensive countermeasures may have to posit, or create, the agents they are intended to counter in order to guide the defensive program or provide opportunities for testing. Moreover, any defensive program may be susceptible to theft of pathogens and materials, or the diversion of information. In the national-security and intelligence communities, this problem is known as espionage, and extensive security and counterintelligence procedures have been developed to counter it. To the extent that defensive research in bioscience is conducted openly, much of the information generated in the course of research is unavoidably available to the targets of that research (see the following section). Although the expansion in biodefense research has been accompanied by stricter controls on access to and use of certain pathogens, the growth in biodefense research budgets, and the increased number of facilities handling these agents and people specifically trained to work with them, have raised opportunities for "insider" access.

A more subtle way in which defensive efforts might stimulate offensive actions is by implicitly legitimizing them. Even though reinforcing the global norm against use of biological weapons remains an important policy objective, investing large amounts of financial and political capital to prepare for an actual attack sends the message that the norm cannot be relied on. On the one hand, potential attackers might conclude that the more prepared society is to counter a biological attack, the less effective such an attack would be and the less interested anyone should be in mounting one. On the other hand, potential attackers might instead conclude that the more willing societies are to invest in biodefenses, the more potent and fearsome biological attacks must be.

Although there is no guarantee that defenses can outrun offenses, most workshop participants believed we have no alternative but to try. Three participants voiced similar views:

"I'm actually not sure the defense can outrun the offense, but to not attempt it makes it even worse. And, historically, while defenses may not have outrun offenses, they've at least advanced the overall state of quality of life." (Robert Erwin)

"I feel that offense…is so easy, and that defense is hard, but that doesn't mean we don't try to mature the defense. We have to do the best we can." (Charles Cantor)

"There is no choice in my view. We have to run faster. We have to figure out what can happen, we have to create the technology to detect what's going on and understand…how whatever shows up works…and we just don't have the tools at hand to go fast enough right now." (Robert Carlson)

Prescriptions for Managing Risks

Under the assumptions that it would be impossible to restrict significantly the overall development of global biotechnology, and that technologies that have already been disseminated are sufficient to pose serious threats, the workshop addressed mechanisms to manage the risk that biotechnology might be turned toward malicious purposes.

Targeted governance mechanisms—those aimed at experiments, at experimenters, or at the dissemination of information and research results—were discussed at the workshop, if not agreed to. Workshop participants struggled with the need to promote legitimate applications of biotechnology, stay aware of state-of-the-art developments, and mount an effective and high-quality defensive research program—all of which require open engagement with the (inherently international) scientific community—while at the same time minimizing the contribution that legitimate scientific activities may inadvertently make to weapons programs.

Regulating Experiments

The Fink Committee and the National Science Advisory Board for Biosecurity

One particular policy measure the U.S. government is instituting today—in response to the recommendation of a National Research Council (NRC) study—attempts to lessen the chance that advanced scientific research might end up facilitating terrorism. The NRC study group, called the Fink Committee after its chair, Gerald Fink from MIT's Whitehead Institute for Biomedical Research, appreciated the importance of preserving the life-sciences community's ability to do research while also acknowledging the science and policy communities' obligations to respond to the risks of misuse of that research.[10]

This study rejected a top-down, government-managed, mandatory approach toward regulating life-science research. Instead, it advocated a voluntary, largely bottom-up process in which proposals for experiments in areas considered to be particularly relevant to biological-weapons development would be reviewed for their potential to produce research results that might be misused. These security reviews would be conducted by local Institutional Biosafety Committees, which are already responsible for ensuring that experiments comply with existing guidance regarding use of recombinant DNA techniques and other biosafety provisions. The Fink Committee also proposed to establish a new national advisory body to provide guidance to these local review boards on how to exercise their new security responsibilities.

To implement these recommendations, the federal government is creating a National Science Advisory Board for Biosecurity (NSABB). The NSABB will oversee a process that would require proposals for federally funded life-science research in certain areas to undergo such reviews. Participation in this process would remain voluntary for nonfederally funded research, but senior federal officials stated their expectation to foster a "culture of responsibility" in which corporations and others performing privately funded research would voluntarily participate in the same process. All those involved in this new process—both on the Fink Committee and in the federal response to the Fink Committee's recommendations—acknowledged that any governance structure for research could only be effective if it were implemented internationally. One of the responsibilities of the NSABB would therefore be to work toward a harmonized, international oversight system.

Both the Fink Committee and the government officials working to implement the NSABB stressed that these mechanisms are not designed to prevent intentional use of biology for harm. Instead, they are intended to lessen the likelihood that legitimate life-science and biotechnology activities would unintentionally contribute to malicious application.

Reaction to the NSABB

Workshop participants generally thought that the Fink Committee recommendations, and the establishment of the NSABB, were steps in the right direction. They agreed that any effective system would have to be international, but they differed on whether it was preferable to start with a U.S.-centered system and attempt to broaden it or to have a more explicitly international framework from the start. In an earlier discussion not explicitly focused on the NSABB, one participant had been quite skeptical about taking a regulatory approach to addressing dual-use concerns, pointing out that people and firms will figure out

---

[10] Committee on Research Standards and Practices to Prevent the Destructive Application of Biotechnology, *Biotechnology Research in an Age of Terrorism.*

ways around any regulation, including moving to countries with less-stringent systems, as the pharmaceutical industry has done.

Some participants expressed concern about the voluntary nature of the proposal review process and the fact that there would be no obligation to include research that was not funded by the government. Others, however, believed that a mandatory system would not be appropriate because legally binding regulations require a specificity that would be very difficult to attain, and because a mandatory system with its associated penalties for noncompliance would threaten the collaborative relationship that the government needs to have with the scientific community to address bioweapons threats effectively.

Workshop participants were also concerned that the NSABB has been precluded from exerting oversight over any classified biodefense research. Elisa Harris found the decision to exclude classified research from the board's scope to be "interesting because, of course, the classified research would presumably be biodefense research, and that was one area that the Fink Committee singled out as posing particular dual-use concerns." Or as another participant put it,

> [Classified biodefense research is] going to be raising the questions that are the diciest…[and] I sure want some people looking at it in ways that the rest of the world can be confident have given it a valid scrub…. It certainly should not be left to individual investigators or individual agencies to wing on their own.

Ultimate Effectiveness of Research Controls

The review process recommended by the Fink Committee can have two qualitatively different objectives. The first objective would be to prevent research with particularly dangerous weapons' implications from becoming available to those who would misuse it. Possible means to preventing its misapplication could include forgoing it completely; restricting the release of information about it; or placing conditions on the people who will perform it (see next section). However, given the strong legitimate incentives to perform life-science R&D, and the necessity for that research to be openly published if it is to become available to those who can make use of it, it is unlikely that very much research that has fundamental scientific merit will end up being rejected or restricted by these review boards. Therefore, it is not clear how much research this new system will effectively keep out of the hands of terrorists.

A second, more modest, objective is to persuade policymakers and the political system that the scientific community is doing everything it can to act responsibly with respect to biological-weapons threats—and that part of acting responsibly is to establish a review process to bring considered, independent judgment to bear on potentially troublesome experiments. The ability to assert that these experiments have been carefully evaluated and judged on balance to be beneficial, and to provide an "audit trail" to document the decision to proceed, may end up being the most important contribution from this new review structure, even if those reviews are unable to prevent research results from becoming accessible to terrorists.

This sort of review structure would have been helpful in the case of certain contentious experiments that have already been performed, such as work to modify modern flu viruses to contain genes from the pandemic 1918 flu that killed millions of people around the world. As Robert Carlson described,

> Fortunately, those strains are directed at mice so this is not something that can immediately jump the species barrier and get out into humans— nonetheless, that work is happening. Let me be clear. I don't think this is a bad idea. We ought to be doing this. However, it's a little scary.

Research that is "scary" to a life scientist might be downright terrifying to a politician, and the absence of any review system could impel the political system to create one. Unfortunately, security mechanisms that are imposed on the scientific community, rather than developed in partnership with it, are likely to damage science without providing much security benefit. Therefore, mechanisms that lessen the chance of such a backlash are important.

Regulating Personnel

In addition to screening proposed research for its potential weapons implications, controls or licensing requirements could be placed on those people whose work could be misapplied to create biological weapons.[11] Even if it did not have much ability to block deliberate malicious activity, such a requirement might nevertheless have some benefit. As Elisa Harris explained,

> If you had the legitimate scientific community organized and adhering to clear procedures and rules, those that stay outside of the system might become a little bit more obvious. People might be more willing to blow the whistle on those that haven't been licensed or have been licensed but are doing things without submitting them to the appropriate level of oversight.[12]

---

[11] Existing U.S. legislation requires that anyone who possesses, uses, or exercises control over "select agents"—those pathogens designated as particularly relevant to potential bioterrorist attack—must be reviewed by the attorney general and specifically approved for access to such agents. These reviews are to ensure that the individuals do not fall into the category of "restricted persons," who are prohibited by law from possessing those agents, and that they are not "reasonably suspected" by federal security agencies of having connections to terrorism. Discussion at the workshop assumed these checks would be performed and primarily addressed whether additional requirements were necessary.

[12] Such an approach has been proposed by the Center for International Security Studies at the University of Maryland. See the working paper by John Steinbruner, Elisa D. Harris, Nancy Gallagher and Stacy Gunther, "Controlling Dangerous Pathogens: A Prototype Protective Oversight System," University of Maryland, September 2003, available at http://www.cissm.umd.edu/documents/Pathogens project monograph_092203.pdf (last accessed February 8, 2005).

While acknowledging there might be some value in such a system, other participants questioned how effective it would be in practice, given the ambiguity of the research itself. "I think if there were people doing bad things, they would have licenses—so I don't think that [requirement is] going to necessarily screen them out," said one. Furthermore, the more extensive and mandatory any licensing scheme becomes, the greater the administrative difficulty in establishing it, and the greater the chance that the process would generate an adversarial, rather than a collaborative, relationship between the scientific and security communities.

Other questions to be addressed regarding licensing of professionals include what groups would require licenses (e.g., researchers, technicians, administrators), the criteria describing which members of those groups would need them, and the implications of imposing licensing requirements either on a very broad segment of the life-sciences and biotechnology workforce, or alternatively requiring them only of individuals working in areas that are designated as particularly relevant to biological weapons and biodefense. In the latter case, given that working on biodefense may have somewhat of a stigma in academic circles that value research on its intrinsic scientific contribution and innovativeness, rather than its relevance to narrow societal objectives, Gigi Kwik, fellow and assistant professor at the University of Pittsburg Medical Center, stated,

> I would be worried about creating a second class of people who are working on biodefense.... I think we need to be concerned more about encouraging scientists to go into biodefense...and creating a second class of citizens, of scientists, doesn't seem like a very good way to get at that goal.

Regulating Materials

The Fink Committee found that safeguarding collections of existing biological agents is a priority that has largely already been met through existing legislation and that the designation of certain pathogens as "select agents" is "an appropriate starting point for identifying strains and isolates that need to be secured."[13] The committee recommended, however, that the NSABB review the security controls applied to biological agents. The NSABB's charter would certainly permit the board to take on such a role, but there is no specific obligation to do so.[14]

Such controls might impede those without specific expertise in selecting and culturing microorganisms from producing pathogens for use as weapons. However, essentially all pathogenic organisms can be extracted from the natural environment. Therefore, the Fink Committee concluded that it is "not feasible—with the possible exception of smallpox—to prevent knowledgeable individuals

---

[13] Committee on Research Standards and Practices to Prevent the Destructive Application of Biotechnology, *Biotechnology Research in an Age of Terrorism*, p. 121.

[14] Charter, National Science Advisory Board for Biosecurity, signed by Tommy Thompson, secretary of health and human services, March 4, 2004. Available at www.biosecurityboard.gov (last accessed February 8, 2005).

from obtaining any of the agents listed on the CDC [Centers for Disease Control] select agent list simply by increasing the physical security of the laboratory environment."[15] Moreover, the many sources for reagents and materials, and the widespread legitimate uses for them, complicate any attempt to regulate bioweapons production through that kind of control.

There might be a role for controls in regulating access to specialized materials that are developed in the process of engineering viruses to perform new functions, an approach that is being explored by a significant number of legitimate laboratories. One example would be the use of plant viruses as a means of genetically engineering plants to produce certain novel products, such as vaccines. Such a process would start with isolation and purification of a virus that naturally infects those plants. The genetic material from those viruses would be purified, analyzed, and modified, possibly with the deletion of genes that cause disease in the host plant and with the addition of genes that code for the desired new product. This modified genetic material would then be reassembled into viable viruses, or vectors, that could carry the desired new gene into the target plants. If everything worked, plants infected by these vectors would remain healthy, but would produce the desired material.

Access to several of the products formed during this process—including the reconstructed genetic material, or the complete viral vectors—would significantly ease the job of anyone who wished to modify that same type of virus for use as a weapon. Therefore, even though these materials themselves might be harmless, controlling their dissemination might help impede biological-weapons production—recognizing that such controls at best would slow down, and could not prevent, anyone who wished to use similar techniques to produce weapons.

Regulating Information

The most difficult policy issue to address, in terms of balancing promotion of legitimate research against the control of illegitimate activities, is that of limiting the dissemination of research information and results. On the one hand, the fact that offense is much easier than defense (i.e., that it is much easier to use bioscience to create weapons than to create countermeasures) argues that effective policy measures to restrain weapons development could be valuable, even if they inhibited somewhat the development of defensive countermeasures or public-health applications. On the other hand, individuals and groups such as the Fink Committee that have looked in detail at this question have had great difficulty in identifying information controls or restraints that would likely prove effective at constraining weapons development without imposing unacceptable costs on legitimate activity. Information describing the design of specific defensive systems, such as the exact genetic sequence that a biodetector looks for to confirm the presence of a pathogen or the sensitivity with which such a device operates, can be kept secret as long as the equipment it refers to is kept secure. However,

---

[15] Committee on Research Standards and Practices to Prevent the Destructive Application of Biotechnology, *Biotechnology Research in an Age of Terrorism*, p. 121.

principles of nature discovered in one place can always be rediscovered somewhere else, no matter how securely the news of the first discovery is kept. Controlling scientific information—even if done effectively—does not eliminate it.

Classification

One set of arguments that is typically made to oppose the use of classification to control biodefense-research information was not discussed at the workshop. The classic argument for keeping biodefense research open and unclassified is that secrecy might be suspected of being used to hide illegitimate offensive activity in violation of the Biological Weapons Convention. Countries that suspect each other of mounting covert offensive programs, according to this line of argument, are more likely to pursue offensive options of their own in response.

These concerns remain valid; yet in today's environment, they must be augmented by the recognition that nonstate terrorist groups may pose as much, or more, of a threat than state weapons programs. These nonstate groups are motivated not by mutual suspicion, but by the deliberate pursuit of mass murder. Complete openness in biodefense research would not only do nothing to moderate their ambitions, it might actually provide them with information that could assist their weapons programs.

Pros and Cons of Restrictions

Does making research information openly available benefit the defense more than controlling such information would inhibit the offense? Even without addressing the mechanics of how a system of reviews and restrictions would be implemented in practice, which would introduce an additional set of difficulties, workshop participants in general concluded that openness was preferable. As one panelist put it, "the more information that's out there, the better." At the same time, they also acknowledged feeling ambivalent at times about disseminating research information that is relevant to bioterrorism.

It is important to recognize that even if information is not published openly, it still might be provided within a restricted environment to those who are in a position to use it. However, there is a cost from failure to publish openly. The more people who are in a position to build on a piece of research, the greater the impact that research will have. Similarly, the more people who are in a position to challenge or validate a reported result, the greater the chance of detecting and eliminating errors. Perhaps most importantly, research results have time and again contributed to breakthroughs by later researchers who would not have been identified at the time of the original work as being likely to need it. These breakthroughs would not have happened had the original research been done under a restricted distribution system that relied on knowing in advance who should receive the results.

Some reasons to support open discussion, even of research results that could benefit offensive weapons programs, include:

- *Staying abreast of technological possibilities.* Publication is essential to alert the scientific, biodefense, and policy communities as to what might be possible. As Robert Carlson stated, "The worst thing we could do in regulating biology is create an opaque black market. The only way we're going to have real insight about what's going on is to keep track of what's going on."

- *Guiding and optimizing defensive efforts.* Public release of research results that identify potential vulnerabilities can indicate areas where defensive countermeasures are needed. Furthermore, exposing defensive research to the widest possible set of colleagues who can validate it or extend it improves the efficiency and quality of the defensive research effort.

- *Informing public debate and motivating responses.* Democracies do not devote resources and attention to problems that their citizens do not know they have. If additional resources will be required to address biological threats, those whose taxes will finance those investments and whose votes elect the leaders who will carry them out need to know what the problem is. Robert Erwin noted, "in some respects, publishing helps to keep the nature of the problem on the front burner and much more highly visible than it would be if there were restraints on publishing."

- *Acknowledging the limited effectiveness of controls.* There are real costs to restraining publication, and some participants did not believe these costs would be worth paying for controls that at best could never be very effective or durable. Under a system of publication restraints, "at best you slow down just a little bit," said one. According to another, "I would tend to come down more on the side of openness than control because of my pessimism that in the long run control works, even though in the short run it might delay dissemination."

- *Avoiding construction of a restriction regime that would be cumbersome at best.* The practical difficulties of managing "top-down" or legally mandated reviews and restrictions were not discussed much at the workshop, but they would pose a formidable challenge. The international scientific-research enterprise is far too decentralized and dynamic for any kind of "central control" to be imposed, even if the legal and even constitutional issues that might preclude such an arrangement entirely could be overcome.[16] Not only would any kind of centralized review constitute a serious and probably unworkable bottleneck, but it would also seem to require either asking those who were not familiar with the paper's technical content to judge its security significance, or asking those who were not familiar with security concerns to address the paper's technical merits. Moreover, given the international nature of science, any effective

---

[16] For example, any system requiring scientists who were not government employees and who were not funded by the government to submit articles for government review raises First Amendment issues. See Elizabeth Rindskopf Parker and Leslie Gielow Jacobs, "Government Controls of Information and Scientific Inquiry," *Biosecurity and Bioterrorism*, vol. 1, no. 2 (2003): 83–95.

review system would have to be imposed internationally—quite an undertaking when just doing so domestically faces so many challenges.

- *Self-restraint.* Given the practical and legal difficulties in imposing a formal, top-down, centrally managed review process, another alternative to unlimited dissemination would be a system in which researchers voluntarily decided either not to write up research results at all, or else to restrict distribution of their publications through a mechanism that has yet to be established. This kind of self-censorship is taking place today.

  Robert Erwin, who opposed a formal policy of restricting publication in general, nevertheless acknowledged that scientists themselves may choose to withhold certain results or findings: "There are cases…where it's pretty easy to lay out certain types of road maps, and scientists have chosen not to do it in conversations with the press or in presentations at public meetings." Elsewhere in the discussion, he revealed that "I personally would not publish a road map to show someone how to engineer a new virus. So I guess my personal decision may be a little different than my inability to come to a strong conclusion on what a public policy ought to be."

  A more extreme category of self-censorship would be the actions that some scientists have reportedly taken to destroy strains of organisms that have been developed for fear that they would be terrible bioweapons if they ever got loose. Gigi Kwik disagreed with this approach, stating that "I'm not sure it's really helping anyone to not examine that strain and see if there's anything we can learn from that for defense. Because if something is important enough or something is easy enough to do, it will be repeated, and then we're going to have to deal with it later." However, exploiting such strains for study, particularly within a classified or restricted environment, might raise questions concerning treaty compliance.

- *The role of private industry.* Most discussions of potential restrictions on research publication assume the issue primarily to affect academic researchers. Although private industry plays a major, if not dominant, role in the evolution of biotechnology today, publication in technical journals is often not as important to industrial researchers as it is to academics. Moreover, if a line of research does not appear to have market potential— as may well be the case for research that could have particular utility for weapons purposes—industrial scientists will be less likely to pursue or to publish it, whereas academic researchers may continue working in that area because of its scientific value. Nevertheless, some workshop participants pointed out that tensions between open communication and restriction are encountered in industry as well, with one reminding that "people who really don't want to publish in the scientific literature—they publish patent literature."

As a model for a self-governance approach within the private sector, one participant raised the Institute of Nuclear Power Operations, an industry-based mechanism that establishes and disseminates best practices among nuclear power-plant operators. Since reactor owners cannot get insurance without being members in good standing, they have a powerful incentive to participate. Another participant mentioned a project, "The Future of the Life Sciences: Reaping the Rewards and Managing the Risks," currently being conducted by two policy research organizations, which is attempting to create such a mechanism in the life sciences community, including industry.[17]

International Mechanisms

The bioweapons threat is international, and international mechanisms will be needed to mitigate it. However, the treaty that most directly applies to biological weapons—the Biological and Toxin Weapons Convention—did not receive much specific discussion at the workshop. This treaty, which promotes the use of biology for peaceful purposes, establishes a clear international norm that biological weapons are illegitimate. However, it has no monitoring provisions, and a multiyear effort to develop a monitoring and compliance protocol for this treaty failed when the United States announced in 2001 that neither the draft under consideration nor any variation of it would be acceptable. This finding was quite controversial at the time but was not revisited at the workshop, which was intended to look forward without getting bogged down rehashing past debates.

An alternate example of an international governance mechanism affecting biotechnology is the Biodiversity Convention. The Biodiversity Convention is intended to help countries protect their native plant and animal species from exploitation, and the convention's Biosafety Protocol is intended to manage any risks to biodiversity posed by the release into the environment of living modified organisms. However, the drafting of this protocol illustrated the difficulty in bringing together the diverse perspectives and sets of expertise that surround any controversial issue, especially one with a significant scientific and technical component. According to one workshop participant who had participated in the development of the Biosafety Protocol, negotiating sessions tended to be dominated by individuals from national environmental ministries who often took positions diametrically opposed to their own ministries of agriculture, economic development, and trade: "A lot of [the countries represented at these negotiation sessions] really want biotechnology to use to improve their agriculture—but the people who want it most aren't at the meeting."

Although the interagency process within the U.S. government by which such internal disputes are mediated is often criticized, said this participant, "at least we

---

[17] This project is being conducted by the Chemical and Biological Arms Control Institute and the International Institute for Strategic Studies—US; further information is available at http://www.cbaci.org/nonp/projects.html (scroll down to the heading "Biotechnology Industry Project") (last accessed February 8, 2005).

have one." Furthermore, Biosafety Protocol meetings faced the additional complication, in this participant's view, of involving "large number of individuals…who know so many things that just aren't true about the risks of biotechnology."

The greater the role that nonstate organizations, rather than states, play in posing biological-weapons threats, the less suited treaties would seem to be as a response. Moreover, treaties have difficulty coping with fields such as biotechnology that are characterized by rapid advances in science and technology. Even so, there is clearly a role for coordinated international activity in countering biological threats. Some workshop participants felt that activities in fora such as the Organization for Economic Cooperation and Development would be more readily agreed to, and more likely to be helpful, than formal treaties or protocols.

Additional Findings and Recommendations

- *Keep some perspective.* According to David Franz, "the bad news [regarding biological threats] is anything is probably possible, but the good news is it's not always as easy for…the ones who might abuse biology as some people might believe."

- *Avoid the Pearl Harbor syndrome,* by which is meant not that we are likely to minimize the threat until we are attacked (which may be true), but that we must avoid the tendency to believe that we will always be able to overcome the initial shock of a surprise attack by mobilizing after the fact. "The problem in biology is we can't wait for the incident and then substantially prepare," said Richard Danzig, "because we won't have a chance to—it runs too fast."[18]

- *Flexibility and rapid response are paramount.* Several participants emphasized the importance of being able to respond rapidly to a biological incident whose nature and probable evolution may be highly uncertain at first.

- *Public awareness of the threat is not very high.* Individuals who chose to spend a day participating in a workshop on the implications of the global evolution of biotechnology for the biological weapons threat are not representative of the attitudes of the public at large regarding bioweapons. Although there was a general consensus in the room that this is a serious problem that deserves attention, that attitude is not necessarily reflected very widely among the general public. When asked whether it was true that many in the international community consider the bioweapons threat to be a "U.S. obsession" and do not take it very seriously themselves, Richard Danzig vigorously disputed the premise of the question. "It's not an American obsession. I wish we were more obsessed with it."

---

[18] See also Danzig, "Catastrophic Bioterrorism."

Appendix. Workshop on Global Evolution of Dual-Use Biotechnology: Biographies of Cited Speakers

Charles Cantor is chief scientific officer and a member of the board of directors, at SEQUENOM, Inc. In addition to founding SelectX Pharmaceuticals, a drug-discovery company, in 2002, he is a member of the board of directors. He is also the director of the Center for Advanced Biotechnology at Boston University and professor of biomedical engineering. Dr. Cantor has held positions at Columbia University and the University of California at Berkeley and was also director of the Human Genome Center of the Department of Energy at Lawrence Berkeley Laboratory. He has published more than 400 peer-reviewed articles, has been granted more than 60 patents, and coauthored a 3-volume textbook on biophysical chemistry and the first textbook on genomics and the Human Genome Project. He sits on the advisory boards of more than 20 national and international organizations and is a member of the National Academy of Sciences.

Rob Carlson is a research scientist in the Department of Electrical Engineering at the University of Washington and was recently a visiting scholar in the Comparative History of Ideas Program. He is also a senior associate at Bio Economic Research Associates (www.bio-era.net). A Princeton-trained physicist, he devotes much of his current research and writing to biology and is working on a book exploring the future of biology as technology. His current research revolves around conceiving and developing new technologies that enable a rapid understanding of biological phenomena and provide a basis for engineering synthetic biological systems. He has codeveloped several patented technologies to detect and quantify small amounts of proteins in complex mixtures and is now working on new fabrication techniques to build microdevices for basic research and clinical diagnostics.

Richard Danzig is a consultant to the Department of Defense on terrorism generally and bioterrorism specifically. He is a director of Human Genome Sciences Corporation, National Semiconductor Corporation, and Saffron Hills Ventures. Mr. Danzig served as the secretary of the navy from November 1998 to January 2001 and currently serves as chairman of the board of the Center for Strategic and Budgetary Assessment and member of the board of directors of Public Agenda and Partnership for Public Service. His academic career includes positions at Stanford University's School of Law and Harvard University's School of Law, and fellowships at the Harvard Society of Fellows and the Rockefeller Foundation. Mr. Danzig received a B.A. degree from Reed College, a J.D. degree from Yale Law School, and B.Phil. and D.Phil. degrees from Oxford University, where he was a Rhodes Scholar.

Robert L. Erwin founded Large Scale Biology Corporation, has served as a director and officer since 1987, and currently serves as chairman of the board. Mr. Erwin is the former chairman of the State of California Breast Cancer Research Council and former board member of the Analytical & Life Science Systems

Association. He is also chairman of the supervisory board of Icon Genetics AG. Mr. Erwin received an M.S. degree in genetics from Louisiana State University. He currently has 6 issued and 14 pending U.S. patents.

David Franz is the chief biological scientist at the Midwest Research Institute and director of the National Agricultural Biosecurity Center at Kansas State University. Dr. Franz served in the U.S. Army Medical Research and Materiel Command for 23 of 27 years on active duty and retired as a colonel. He has served as commander of the U.S. Army Medical Research Institute of Infectious Diseases (USAMRIID) and as deputy commander of the Medical Research and Materiel Command. Dr. Franz was the chief inspector on three UN Special Commission biological-warfare inspection missions to Iraq and served as technical adviser on long-term monitoring. He also served as a member of the first two U.S.-UK teams that visited Russia in support of the Trilateral Joint Statement on Biological Weapons and as a member of the Trilateral Experts Committee for biological weapons negotiations. Dr. Franz is a resident graduate of the Army Command and General Staff College, a recipient of the Army Research and Development Achievement Award, the Order of Military Medical Merit, and the Legion of Merit with oak leaf cluster. He holds a D.V.M. from Kansas State University and a Ph.D. in physiology from Baylor College of Medicine.

Elisa Harris is a senior research scholar at the University of Maryland's Center for International and Strategic Studies. In the past, she served as director for nonproliferation and export controls on the National Security Council staff (1993-2001), where she was responsible for coordinating U.S. policy on chemical, biological, and missile proliferation issues. She has held a number of research positions, including the Brookings Institution, the Royal United Services Institute for Defense Studies in London, and the Center for Science and International Affairs at Harvard University. She is a former SSRC-MacArthur Foundation fellow in international peace and security studies and staff consultant to the Committee on Foreign Affairs, U.S. House of Representatives. Ms. Harris is the author of numerous publications on chemical and biological weapons issues and has testified frequently on such issues before the U.S. Congress. She holds an A.B. in government from Georgetown University and an M.Phil. in international relations from Oxford University.

Gigi Kwik, an immunologist and policy analyst, is a fellow at the Biosecurity Center of University of Pittsburgh at UPMC and assistant professor at the University of Pittsburgh. She is associate editor of *Biosecurity and Bioterrorism: The Journal of Biodefense Strategy, Science, and Practice*, as well as lead researcher on a Nuclear Threat Initiative–funded project on malevolent uses of biotechnology. Dr. Kwik was a founding member of the UPMC Biosecurity Center and, prior to joining the faculty there in 2003, was a fellow and assistant scientist at the Johns Hopkins University Center for Civilian Biodefense Strategies, which she joined in 2001. From 2000 to 2001 Dr. Kwik was a National Research Council postdoctoral associate at the U.S. Army Medical Research Institute of Infectious Diseases (USAMRIID) at Fort Detrick, Maryland.

Kris Venkat is currently chairman of Morphochem, Inc., Transvivo Inc., and Automated Cell, Inc. In addition, he serves as chairman of the supervisory boards of Accentua Pharma AG and Juelich Enzyme Products GmbH, both based in Germany. He has over 25 years of broad experience with both large corporations, including Merck & Co. and H J Heinz Co., and start-up companies. Dr. Venkat has been associated with numerous high-technology companies as a founder, board member, or adviser, including Sequenom and Celsion, Genscope, Synosis, Filtron, Goosecross Cellars, LifetecNet, and Nijjer Agro Foods. He is also a senior investment adviser to Techno Ventures Management, one of Europe's largest venture capital funds. Dr. Venkat serves as visiting professor of biochemical engineering at Rutgers University. He has held visiting faculty positions at Yale University, Dartmouth College, Anna University (India), and University College, Galway (Ireland). He received his undergraduate education at the Indian Institute of Technology in Madras and his Ph.D. from Rutgers University.

About the Author

Gerald L. Epstein is senior fellow for science and security in the CSIS Homeland Security Program, where he is working on issues including reducing and countering biological-weapons threats, bridging the scientific research and national security communities, and examining the role of technology in homeland security. He is also an adjunct professor in the Security Studies Program at Georgetown University's Edmund A. Walsh School of Foreign Service. Previously, Dr. Epstein was with the Institute for Defense Analyses (IDA), prior to which he served with the White House Office of Science and Technology Policy. From 1983 to 1989 and again from 1991 until its demise in 1995, he was at the Congressional Office of Technology Assessment, where he worked on international security topics. A fellow of the American Physical Society and a member of the editorial board for the journal *Biosecurity and Bioterrorism*, he is a coauthor of *Beyond Spinoff: Military and Commercial Technologies in a Changing World* (Harvard Business School Press, 1992). He received S.B. degrees in physics and electrical engineering from MIT and a Ph.D. in physics from the University of California at Berkeley.